Whiz-Kid MATHS
WORKBOOK
5 - 7 Years

Numeracy for all - With answers

Addition
Subtraction
Division
Multiplication
Colouring
Codes
Shapes
Reasoning Skills

Best Start For Early Success!

Arinze Oranye

AWARD WINNING AUTHOR

Whiz-Kid Maths Workbook 5 – 7 Years

Whiz-Kid Maths Workbook 5 – 7 years is designed to encourage children in the areas of number work, puzzles, time, transformations, shape, space and measure, basic algebra, general numeracy skills using mathematical concepts.

There are some harder topics, but that would challenge the gifted and talented children.

This book is a household item for early numeracy adaptation necessary for future engagement in mathematics. It is an excellent revision package for KS1 pupils including full answers, test paper and lots of practice.

© 2017 ARINZE EDWARD ORANYE. All Rights Reserved.

No part of this book may be reproduced, stored in a retrieval system, or transmitted by any means, electronic or mechanical, photocopying, recording or otherwise without the written permission of the author.

This book may not be lent, resold, hired out or otherwise disposed of by the way of trade in any form other than that in which it is published, without the written permission of the author.

To order this workbook in bulk, contact Iconic Concepts Limited by phone on +44 07825313771, or by email: edwardoranye@hotmail.com, edward@iconicconcepts.co.uk.
Website: www.iconicconcepts.co.uk

CONTENTS

The Workbook	4	Reflections	35
Numeracy	10	Dividing Whole Numbers	36
Colouring Activity (1)	12	Colouring Activity (3)	37
Words and Numbers	14	Lengths and estimates	38
Even and Odd Numbers	15	Reading scales	39
Adding Smaller Numbers	16	Frequency table	40
Place Value	17	Bar charts	40
Missing Numbers	18	Using Codes	41
Find My Number	19	Test	42
Colouring Activity (2)	20	Answers	44
Small and Big Numbers	21		
Number Patterns	22		
2D Shapes	24		
3D Shapes	25		
Symbols and Equations	26		
Thinking Skills	27		
Addition skills	28		
Subtraction skills	29		
Ordering whole numbers	30		
Time and clock 1	31		
Time and clock 2	32		
Number pyramid	33		
Symmetry	34		

THE WORKBOOK

This workbook covers the following topics:

- Place Value
- Colouring activities
- Time and Clock
- Addition and subtraction
- Multiplication and division
- General Arithmetic

PLACE VALUE

The place occupied by a figure affects its value as we shall find out shortly.

In the number **567**, the digits have different values since they take up different places in the order of events. Using the place value table, the number 567 takes up different positions as shown:

Formerly called **units**

Hundreds	Tens	Ones
5	6	7

The value of the digit 5 is five hundred or **500**
The value of the digit 6 is 6 tens or **60**

The value of the digit 7 is 7 ones or simply, **7**
To explain the composition of the number 567, we say that it is made up of five hundred, six tens, and seven ones.

Example 1:
4 2   the 4, has value **40**

5 **1**   the 1, has value **1**

5 **8** 6   the 8, has value **80**
7 1 6   the 7, has value **700**

ADDITION OF WHOLE NUMBERS

There are different ways of adding whole numbers. You can either work out calculations mentally (in your head), work it out on paper or use a calculator which is the easier method.

However, you will not be able to use a calculator all the time as some methods of assessment(s) do not require the utilisation of a calculator.

METHOD 1	METHOD 2
$72 + 63$ $= 70 + 2 + 60 + 3$ $= 70 + 60 + 2 + 3$ $= 135$	$72 + 63$ $7 + 6 = 13 \quad \begin{array}{r} 72 \\ + 63 \\ \hline 135 \end{array} \quad 2 + 3 = 5$
$465 + 183$ $= 400 + 60 + 5 + 100 + 80 + 3$ $= 400 + 100 + 60 + 80 + 5 + 3$ $= 648$	$465 + 183$ $4+1 = 5 \quad \begin{array}{r} 465 \\ + 183 \\ \hline 648 \\ 1 \end{array} \quad 5 + 3 = 8$ $5+1 = 6 \quad\quad\quad\quad 6 + 8 = 14$
$3865 + 129$ $= 3000 + 800 + 60 + 5 + 100 + 20 + 9$ $= 3000 + 800 + 100 + 60 + 20 + 5 + 9$ $= 3994$	$3865 + 129$ $\begin{array}{r} 3865 \\ + 0129 \\ \hline 3994 \end{array}$

Add $343 + 19$

$$\begin{array}{r} 343 \\ + 19 \\ \hline 362 \\ 1 \end{array} \checkmark$$

SUBTRACTING WHOLE NUMBERS
Some calculations can be done mentally.
Examples
$6 - 2 = 4 \quad\quad\quad 100 - 8 = 92$
$10 - 7 = 3$
When subtractions are difficult to do mentally, we can do them on paper.

Example 1: 486 – 43

```
    4   8   6
-   0   4   3
   ─────────
    4   4   3
```

> Always remember to start subtracting from the right side (ones end)
>
> 6 – 3 = 3
> 8 – 4 = 4
> 4 – 0 = 4

Example 2: 573 – 19

```
        6   1
    5   7̷   3
-   0   1   9
   ─────────
    5   5   4
```

> 3 is smaller than 9, so we take 1 from the next column (tens) and 7 reduces to 6. Place that 1 in front of 3 to make 13.
>
> 13 – 9 = 4
> 6 – 1 = 5
> 5 – 0 = 5

Example 3: 800 - 253

```
    7   9   1
    8̷   0̷   0
-   2   5   3
   ─────────
    5   4   7
```

MULTIPLYING WHOLE NUMBERS

Some multiplications can be done straight away in your head. However, some cannot.

$$\left.\begin{array}{l} 5 \times 3 = 15 \\ 8 \times 2 = 16 \\ 4 \times 9 = 36 \\ 7 \times 7 = 49 \\ 10 \times 10 = 100 \end{array}\right\} \text{can be done mentally}$$

257 × 389 needs to be worked out on paper unless you are a genius or a walking computer. Multiplication table can help when multiplying whole numbers.

TABLE 2A

×	1	2	3	4	5	6	7	8	9	10
1	1	2	3	4	5	6	7	8	9	10
2	2	4	6	8	10	12	14	16	18	20
3	3	6	9	12	15	18	21	24	27	30
4	4	8	(12)	16	20	24	28	32	36	40
5	5	10	15	20	25	30	35	40	45	50
6	6	12	18	24	30	36	42	48	54	60
7	7	14	21	28	35	42	49	56	63	70
8	8	(16)	24	32	40	48	56	64	72	80
9	9	18	27	36	45	54	(63)	72	81	90
10	10	20	30	40	50	60	70	80	90	100

2 × 8 = 16   7 × 9 = 63

Three numbers can also be multiplied together.

1 × 3 × 2 = 6

2 × 3 × 4 = 24

1 × 2 × 7 = 14

Example 1: 13 × 4

```
   13
×   4
   52
    ①
```

Always start your multiplication from the ones column.
3 × 4 = 12
Write down the digit **2** in the ones column and move the **1** over to the tens column. 4 × 1 = 1 plus the 1 (moved over) = 5.

Example 2: 24 × 12

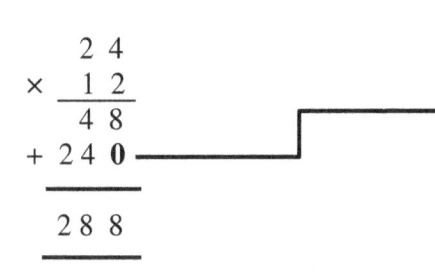

2 × 4 = 8
2 × 2 = 4

Add zero before multiplying by 1

1 × 4 = 4
1 × 2 = 2
Therefore, 24 × 12 = **288**

DIVIDING WHOLE NUMBERS

Division is the opposite of multiplication.

Example 1: $6 \div 3$

> $6 \div 3$ simply means how many times 3 go into 6. In this case, the answer is 2

Example 2: $48 \div 12$

> Knowing your times table helps in division. $48 \div 12$ means how many times 12 go into 48. In this case, the answer is 4.

WORKBOOK →

NUMERACY

Fill in the missing numbers.

1) 1 + 5 =

2) 3 + 6 =

3) 6 + 0 =

4) 10 + 8 =

5) 7 + 3 =

6) 4 + ☐ = 7

7) ☐ + 3 = 9

8) 3 + 5 =

9) 9 + ☐ = 11

10) 7 + ☐ = 8

11) Look at the numbers below.

11   4   5   1   20   17   8

a) Write the even numbers.

b) Write the odd numbers.

c) Write all the numbers greater than 8.

d) Write all the numbers less than 11.

e) What is the lowest number?

f) What is the highest number?

g) Add 4 and 17.

h) Subtract 11 from 20.

i) Work out the value of 5 + 11

j) Add up all the odd numbers.

k) Add up all the even numbers.

Fill in the missing numbers.

12) 9 − 2 = ☐

13) 16 − 5 = ☐

14) 13 − 4 = ☐

15) 25 − 13 = ☐

16) 8 − ☐ = 3

17) 19 − ☐ = 5

18) ☐ − 4 = 13

19) ☐ − 6 = 21

20) 27 − 12 = ☐

21) Perform the calculations and complete the missing numbers.

+	3	5		8
9				
		8	13	

22) Three rectangles are joined together as shown.

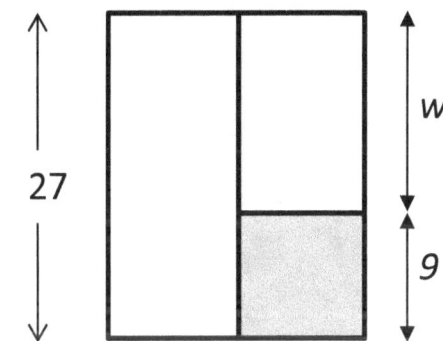

a) Work out the length w. ☐

b) Fill in the missing numbers.

i) 9 + w = ☐

ii) ☐ − ☐ = w

COLOURING ACTIVITY 1

Work out the answers to each question and by using the key, colour each diagram on the right.

Key: 10 = Red   7 = Blue   4 = Green
5 = Brown   20 = purple

1) $3 + 7 =$ 10

2) $7 - 3 =$

3) $13 + 7 =$

4) $12 + 8 =$

5) $10 - 6 =$

6) $5 + 0 =$

7) $2 + 5 =$

8) $9 - 4 =$

9) $13 - 8 =$

10) $0 + 7 =$

11) $4 + 6 =$

12) $17 - 13 =$

MULTIPLICATION

1) 0 x 1 =
2) 1 x 2 =
3) 1 x 3 =
4) 2 x 7 =
5) 2 x 9 =
6) 3 x 0 =
7) 3 x 7 =
8) 4 x 8 =
9) 4 x 9 =
10) 5 x 0 =

11) 6 x 2 =
12) 6 x 4 =
13) 7 x 6 =
14) 7 x 5 =
15) 8 x 2 =
16) 8 x 4 =
17) 9 x 2 =
18) 9 x 5 =
19) 10 x 3 =
20) 10 x 10 =

Work out the value of the numbers under the smiley faces.

21) ☺ x 3 = 6
22) 1 x ☺ = 2
23) ☺ x 7 = 14
24) 2 x ☺ = 10

25) 2 x ☺ = 8
26) 9 x ☺ = 18
27) ☺ x 7 = 7
28) ☺ x 1 = 8

WORDS AND NUMBERS

Write these numbers in words.

1) 1 ▭
2) 5 ▭
3) 10 ▭
4) 17 ▭
5) 20 ▭

6) Using a ruler, join each number to its word.

5   7   10   15   20

Ten   Five   Twenty   Seven   Fifteen

Write these words in numbers.

7) Eight ▭
8) Twelve ▭
9) Twenty ▭
10) Thirty ▭
11) Thirty – five ▭
12) Fifty – three ▭
13) Sixty ▭

14) sixty – six ▭
15) Seventy ▭
16) Seventy – nine ▭
17) Eighty ▭
18) Eighty – seven ▭
19) Ninety ▭
20) One hundred ▭

EVEN AND ODD NUMBERS

1 From the numbers in the cloud, tick the even numbers.

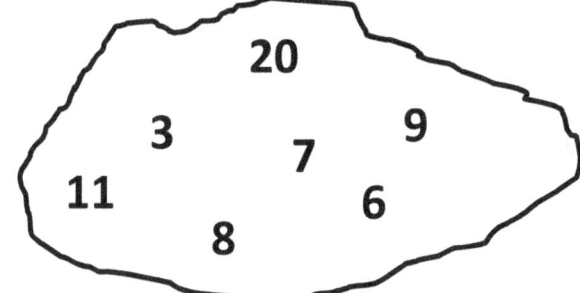

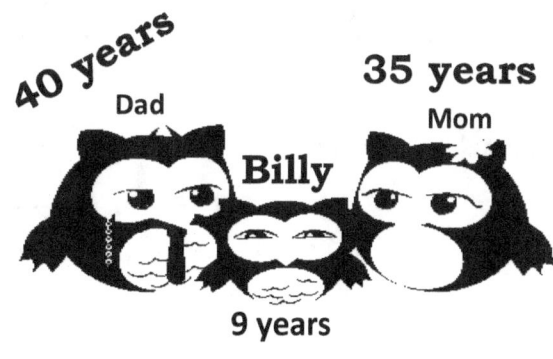

5) Whose age is even?

6) Altogether Billy, Mom and Daddy's age is

2 From the list of numbers above, list all the odd numbers.

3 Write down all the even numbers between 19 and 31.

7 The answer to question 6 above is even. True or false?

8 Dad is older than Billy by [   ] years.

9 The answer to question 8 above is even. True or false?

4 Look at the numbers below:

3, 5, 6, 10, 13, 20

a) Add all the even numbers.

b) Add all the odd numbers.

c) Write down the **difference** between the total of all the even numbers and the odd numbers.

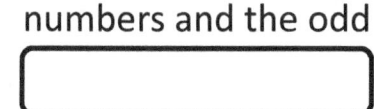

10 Colour all even numbers **red** and all odd numbers **blue**.

1	2	3	4	5	6	7	8
9	10	11	12	13	14	15	16
17	18	19	20	21	22	23	24
25	26	27	28	29	30	31	32
33	34	35	36	37	38	39	40
41	42	43	44	45	46	47	48
49	50	51	52	53	54	55	56

ADDING WHOLE NUMBERS

1) Add the following

a) 12 + 8 = ☐

b) 11 + 9 = ☐

c) 7 + 18 = ☐

d) 12 + 0 = ☐

2) Add the following

a) 11 + 12 = ☐

b) 13 + 15 = ☐

c) 126 + 17 = ☐

d) 211 + 61 = ☐

3 Fill in the boxes below.

a) 12 + ☐ = 20

b) 14 + ☐ = 29

c) ☐ + 7 = 19

d) ☐ + 9 = 15

e) 0 + 11 = ☐

f) 18 + ☐ = 27

g) 20 + ☐ = 50

h) 40 + ☐ = 50

i) ☐ + 23 = 40

j) ☐ + 10 = 33

k) 0 + 50 = ☐

l) 5 + ☐ = 45

4) 8 Penguins were found in a river. **12** more were added to the river. How many Penguins altogether are there in the river? ☐

5) Isabel wants to buy a ruler from a shop. The cost of one ruler is **20 pence**. She bought **3** rulers.
   a) How much did Isabel pay for the three rulers? ☐
   b) Isabel paid for the rulers with a **£1** coin. How much change would she get? ☐

PLACE VALUE

Write down the value of the **underlined** numbers.

1) <u>1</u> 3

2) <u>1</u> 7

3) 2 <u>4</u>

4) 2 <u>9</u>

5) <u>3</u> 0

6) 3 <u>7</u>

7) 4 <u>2</u>

8) <u>4</u> 9

9) <u>5</u> 3

10) 5 <u>6</u>

11) 6 <u>2</u>

12) <u>6</u> 5

13) <u>7</u> 3

14) 7 <u>8</u>

15) 9 <u>7</u>

Write down the value of the numbers in **bold**.

16) 1 **2** 3

17) **1** 3 0

18) **2** 3 4

19) 2 7 **8**

20) **3** 0 9

21) 3 **5** 7

22) 4 **2** 8

23) 4 9 **5**

24) **5** 1 3

25) 5 6 **7**

26) 6 **1** 4

27) 6 8 **2**

28) **7** 3 1

29) 7 **6** 5

30) **8** 2 9 4

MISSING NUMBERS

If all the numbers are arranged in order, fill in the missing numbers.

1) 2, 3, 4, ☐, ☐, ☐

2) 5, 10, ☐, 20, 25, ☐

3) 11, ☐, 15, ☐, 19, 21, 23

4) 5, 8, 11, ☐, ☐, 20,

5) ☐, 3, 6, 9, ☐, ☐

6) 5, 10, 15, ☐, ☐, ☐

7) 10, ☐, ☐, 40, 50, ☐

8) 4, ☐, 12, 16, ☐

9) 1, 4, ☐, 10, ☐, 16

10) ☐, ☐, ☐, 40, 42, 44

11) ☐, 24, 28, ☐, 36

12) 0, 3, ☐, ☐, 12, 15

13) ☐, ☐, 11, 16, 21, ☐

14) 14, ☐, 18, ☐, 22

15) ☐, 12, 15, 18, ☐

16) ☐, ☐, ☐, 20, 24, 28

17) Rearrange the numbers below to make the smallest and largest numbers.

Numbers	Smallest	Largest
13		
25		
69		
213		
328		
304		
423		
439		
527		
132		
743		
834		
511		
781		

FIND MY NUMBERS

Fill in the boxes by **adding** or **subtracting** the given numbers.

1)

	2	8	3	4	7
		3		30	
Total	9		15		20

2) Write different ways to make 40 by completing A and B using addition **or** subtraction.

A	B	40
		40
		40
		40
		40
		40

3) Write different ways to make 57 by completing C and D using addition **or** subtraction.

C	D	57
		57
		57
		57
		57
		57
		57
		57

COLOURING ACTIVITY 2

1) Work out the calculations from the grid below.
2) Colour your answer using the key:

0 – BROWN	3 – RED	6 - WHITE
1 – GREEN	4 – BLACK	10 - PURPLE
2 – YELLOW	5 – PINK	20 - BLUE

12 - 12	3 + 1	1 + 9	6 - 2	12 - 9	19 + 1	10 - 5	16 – 15	5 – 3	7 - 4	5 + 5
0 + 1	2 + 0	3 + 7	6 – 3	1 + 3	7 + 13	9 - 4	11 - 8	40 – 30	2 × 5	1 + 4
8 - 5	12 - 7	7 - 1	12 - 6	10 - 0	12 + 8	6 × 1	0 + 2	5 × 1	10 - 10	1 × 10
2 × 2	1 + 4	0 × 12	43 - 37	1 × 5	25 - 5	8 – 6	9 - 6	13 - 8	6 + 4	9 - 3
4 + 2	7 + 3	11 - 7	3 + 3	11 - 8	9 + 11	2 + 8	5 - 5	15 - 5	6 + 4	7 - 6
9 - 5	34 - 31	11 - 11	1 + 3	8 - 5	30 - 10	90 - 86	50 - 40	8 - 4	17 - 11	20 - 12
10 × 2	7 + 13	40 – 20	4 ×5	6 + 12	40 - 20	11 + 9	16 + 4	50 – 30	8 + 12	20 + 0
9 - 8	67 × 0	18 - 13	8 - 4	11 - 9	23 - 3	5 × 1	4 + 0	40 - 30	17 - 12	51 - 50
0 + 1	8 - 5	9 - 7	0 + 0	10 - 9	8 + 12	89 - 88	15 - 13	5 × 2	10 - 5	90 - 88
5 × 1	1 + 1	1 + 2	2 - 1	8 + 2	5 + 15	1 + 3	17 - 16	76 - 71	3 - 2	14 - 12
3 + 0	50 - 30	3 × 0	12 - 9	3 × 0	0 + 20	9 - 4	9 – 9	9 + 1	15 - 5	1 + 9
11 + 9	5 × 4	16 + 4	2 × 1	15 - 9	16 + 4	6 - 3	0 + 4	30 - 27	65 - 5	30 - 30
0 + 6	10 + 10	7 - 4	9 - 5	4 + 6	10 + 10	1 + 3	40 - 34	3 + 2	43 – 40	50 - 40

SMALL AND BIG NUMBERS

From the pairs of numbers below, circle the **smaller** number.

1) 12   21
2) 13   1
3) 26   14
4) 35   37

From the pairs of numbers below, circle the **bigger** number.

5) 4   11
6) 40   39
7) 50   40

Put a tick on the **smaller** number from each pair.

8) 3.4   5.4
9) 8.1   8.0
10) 1.1 1.7

11) 12.5   21.5
12) 5.5   4.4
13) 34.7   34.4

21

NUMBER PATTERNS (SEQUENCES)

1) Fill in the missing numbers

a) 1, 2, 3, 4, ☐ , ☐

b) 2, 4, 6, 8, ☐ , ☐

c) 5, 10, 15, 20, ☐ , ☐

d) 1, 3, 5, 7, 9, 11, ☐ , ☐

e) 20, 18, 16, 14, ☐ , ☐

2) Write the missing numbers for the sequences below.

a) 4, 5, 6, ____, 8, ____, 10

b) 10, ____, 30, 40, ____, 60

c) 7, 10, ____, 16, 19, ____, 25

d) 1, 5, 9, ____, 17, ____

e) 72, 74, 76, 78, ____ , ____

★	☺	cylinder	pentagon	cross	cube
6	10	14	☐	☐	☐

3) Fill in the missing boxes for the shapes numbers above.

4) Look at question 3 above. What word (Even or Odd) describes the numbers in the sequence? ☐

5) The rule for this sequence 3, 5, 7, 9, 11........ is **ADD 2**
Write down the **rule** for each sequence below.

a) 1, 4, 7, 10, 13,..... ☐

b) 10, 12, 14, 16, ... ☐

c) 20, 17, 14, 11, ... ☐

d) 2, 7, 12, 17, ... ☐

6)

Pattern Position: 1 (○)   2 (○○○)   3 (○○○○○)

a) Draw the next pattern in the sequence. [ ]

b) How many circles are added each time? [ ]

c) Write down the **rule** for the pattern sequence. [ ]

7)

Sequence	3	8	13		23	
Pattern position	1	2	3	4	5	6

a) Complete the table above
b) Write down the **rule** for the sequence [ ]

8) Look at the number sequence below:

40   35   30   25   20

What is the rule for the sequence? [ ]

9 Write down the next **3** terms of the sequences below.

a) First number is 1. Rule: add 2. [ ]

b) First number is 4. Rule: add 3. [ ]

c) First number is 10. Rule: subtract 2. [ ]

d) First number is 17. Rule: Subtract 3. [ ]

2D SHAPES

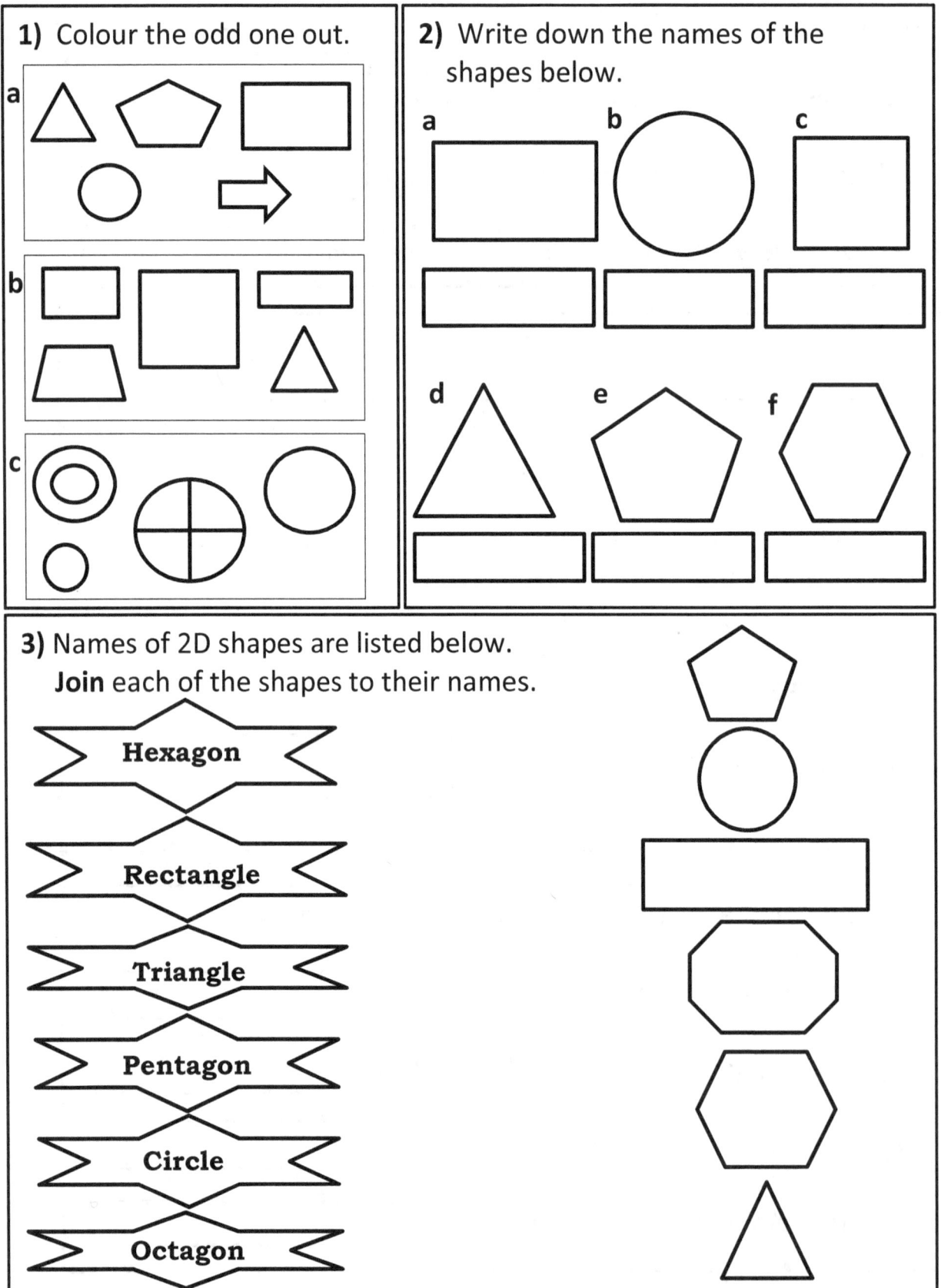

3D SHAPES

1) Write the names of the 3-D shapes below.

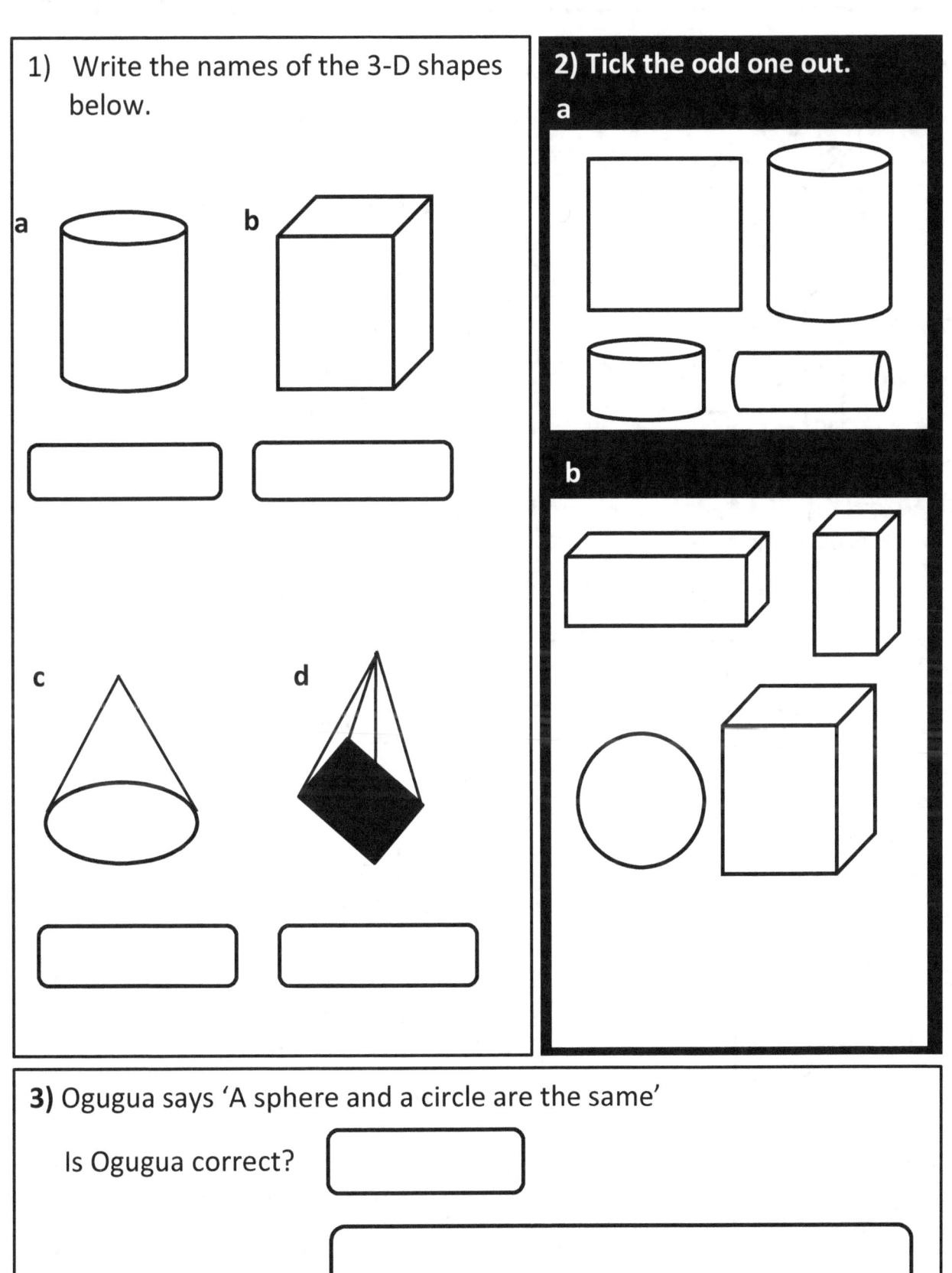

2) Tick the odd one out.

3) Ogugua says 'A sphere and a circle are the same'

Is Ogugua correct?

Explain your answer.

SYMBOLS AND BASIC EQUATIONS

1) Work out the value of the symbols in each equation.

a) $3 + \square = 9$
 $\square = \square$

b) $10 + \square = 20$
 $\square = \square$

c) $1 + \square = 14$
 $\square = \square$

d) $\star + 7 = 20$
 $\star = \square$

e) $\star + 8 = 17$
 $\star = \square$

f) $\star + 25 = 50$
 $\star = \square$

2 Write down the value of each symbol.

a) $\square + 3 = 10$
 $\square = \square$

b) $\bigcirc + 6 = 6$
 $\bigcirc = \square$

c) $\diamond + 5 = 20$
 $\diamond = \square$

d) $\heartsuit - 5 = 15$
 $\heartsuit = \square$

e) $\triangleright - 7 = 13$
 $\triangleright = \square$

f) $\star + \star = 40$
 $\star = \square$

g) $\updownarrow + 100 = 150$
 $\updownarrow = \square$

THINKING SKILLS

1 Reece bought four cups for £2.
Complete the sentences below.
   a) One cup costs _____

   b) Two cups costs _____

   c) Eight cups costs _____

2 A ruler costs 70 pence. Isabel pays with a £1 coin.

   a) How much change does she get? _____

   b) James bought two rulers. How much did he spend?

   c) Rosie had £3. She wants to buy 4 rulers. Does she have enough money? _____

3 Complete the missing numbers.
   a) 2 + 7 = 1 + _____

   b) 20 − 4 = 4 × _____

4 Amber bought this picture and gave the shop assistant £25.

a) How much **more** will she pay? _____

b) Complete the boxes below for the amount of coins(pence) that will complete the payment

   i) 20p + 20p + 10p + 5p + ☐

   ii) 10p + 20p + ☐ + ☐ + 5p

   iii) ☐ + 50p + ☐

c) Tom wants to buy the same picture. He gives £25.60. How much change will he get? _____

ADDITION SKILLS

Example 42 + 53

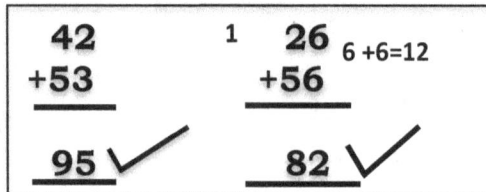

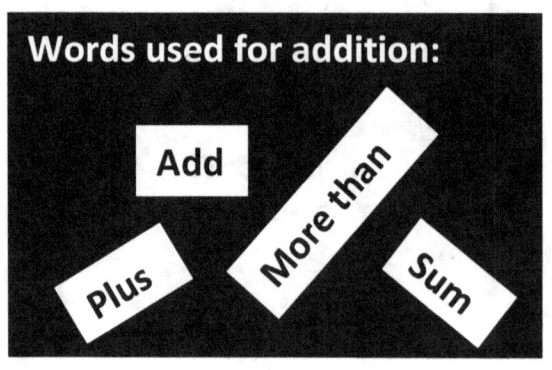

Words used for addition:
Add, Plus, More than, Sum

1. Work out these additions.

   a. 15 + 32   b. 24 + 31

   c. 23 + 53   d. 76 + 22

   e. 55 + 42   f. 24 + 64

2. Work these out.

   a. 13 + 7 = …………

   b. 30 + 50 = …………………

   c. 300 + 500 = …………………

3. What is the sum of 33 and 15?

4. Add 12 and 83

5. 40 plus 50

6. What is 20 more than 200?

Add the numbers below:

7. 76 + 10 =

8. 48 + 30 =

9. 24 + 20 =

10. 71 + 23 =

11. 99 + 0 =

SUBTRACTION SKILLS

35 −13	94 - 43
35 − 13 ――― 22 ✓	94 − 43 ――― 51 ✓

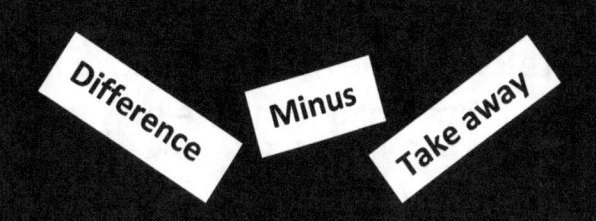

Words used for subtraction.

Difference   Minus   Take away

Subtract these numbers.

1. 65 − 4

2. 29 - 7

3. 68 − 16

4. 44 - 10

5. 87 − 5

6. 71 - 30

7. Work out the difference between 16 and 6 ☐

8. 54 minus 44 = ☐

9. 67 take away 13 = ☐

Function Machines

Work out the missing numbers for **'IN'** or **'OUT'** in the function machines below.

10. 29 → -9 → OUT ☐

11. 20 → -13 → OUT ☐

12. IN ☐ → -3 → 5

ORDERING WHOLE NUMBERS

Ordering simply means placing numbers in order of size from smallest to highest of highest to smallest.

4 is **bigger than** 2, 20 is **smaller than** 30 and 70 is **equal to** 70

< Represents smaller than
> Represents bigger than
= represents equal to

Therefore, 2 is less than 3 and we may write as **2 < 3**
10 is greater than 7 and we may write as **10 > 7**

Write each number in order of size, smallest first.

1. 4, 3, 8, 6, 10, ………, ………, ………, ………, …………

2. 19, 10, 8, 3, 9, ………, ………, ………, ………, ………

3. 20, 10, 6, 4, 16, ………, ………, ………, ………, ………

4. 41, 14, 40, 56, 24, ………, ………, ………, ………, ………

5. 50, 56, 20, 79, 76, ………, ………, ………, ………, ………

Write the missing **signs** for each pair of numbers with <, > or =.

6. 5 ☐ 8   **7.** 30 ☐ 67

8. 10 ☐ 12   **9.** 80 ☐ 20

10. 20 ☐ 10   **11.** 21 ☐ 47

12. 7 ☐ 70   **13.** 200 ☐ 199

Time and Clock 1

This clock shows 2 o'clock

What time is shown on these clocks?

1. 
2. 
3. 
4. 
5. 
6. 
7. 
8. 
9. 
10. 
11. 
12. 
13. 
14. 
15. 
16. 
17. 
18. 
19. 
20. 
21. 
22. 
23. 
24.

TIME AND CLOCK 2

1) Draw the hands of these clocks to show the times below each clock.

6 o'clock   **12 o'clock**   **9 o'clock**   **1 o'clock**

2) Write the missing clock numbers on the clock face below.

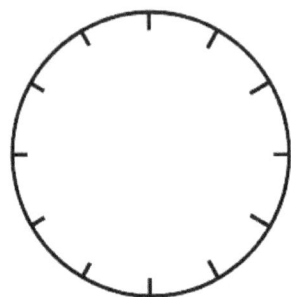

3) Draw the hands of these clocks to show the times below each clock.

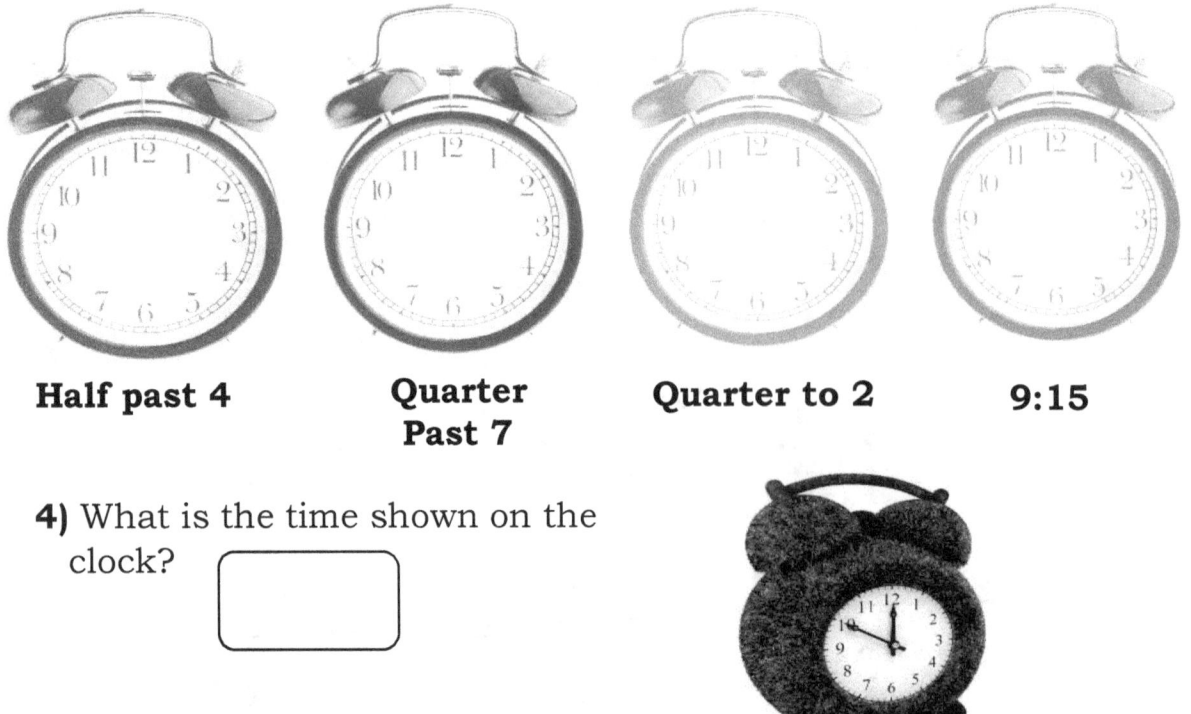

Half past 4   **Quarter Past 7**   **Quarter to 2**   **9:15**

4) What is the time shown on the clock?

32

NUMBER PYRAMIDS

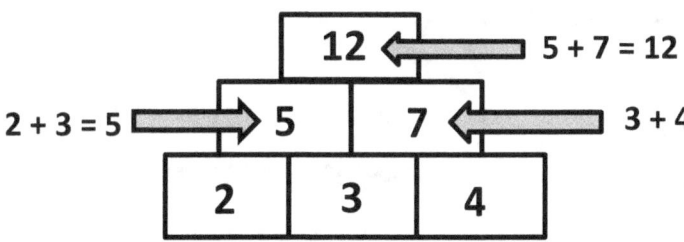

1) Look at the example above. Complete each pyramid below.

The two bottom numbers are added to give the number on top

a 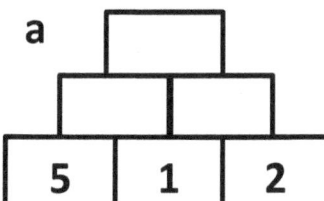   b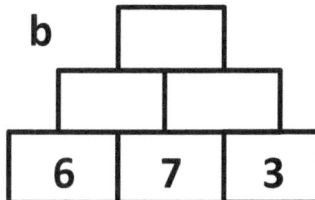

2) The diagram below is a **multiplication** pyramid. Complete the boxes.

c 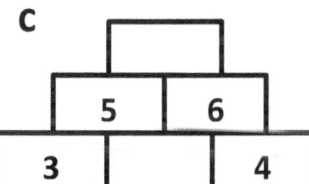   d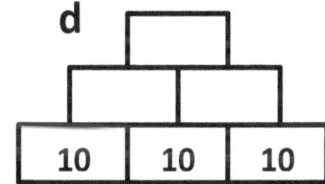

a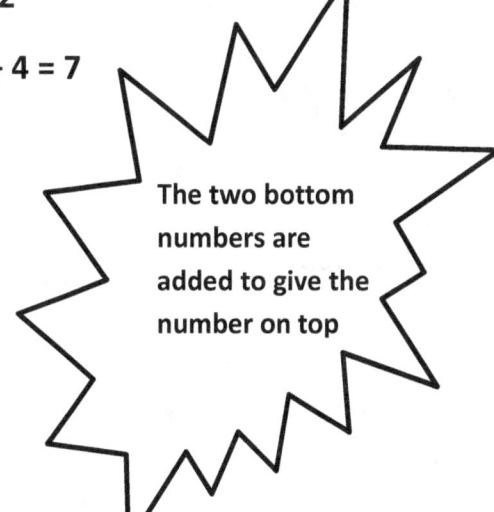

e 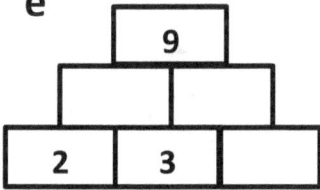   f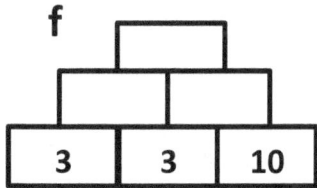

b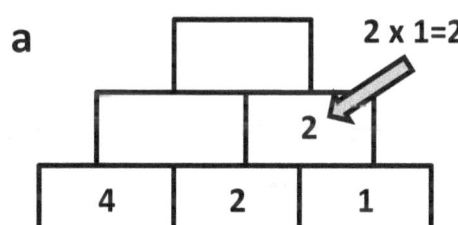

g 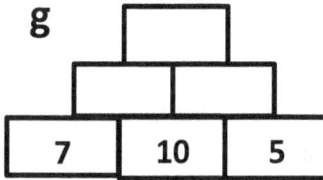   h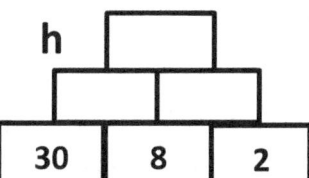

c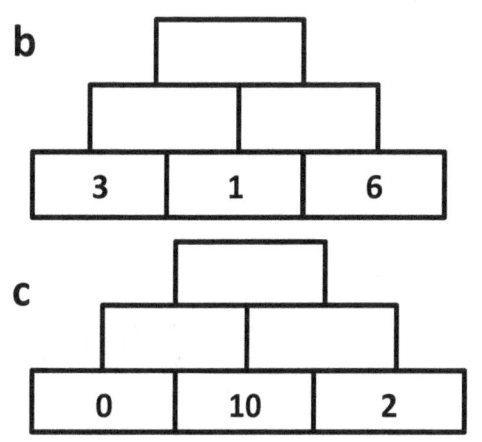

SYMMETRY

If we fold these shapes along the dotted line, one half of the shape will **fit exactly** on top of the other half.

Some shapes may have more than one line of symmetry

Line of symmetry

The dotted line is called the **line of symmetry**. It is the line that divides the shape into **two** identical halves.

1 Complete the table and draw the line(s) of symmetry on each shape.

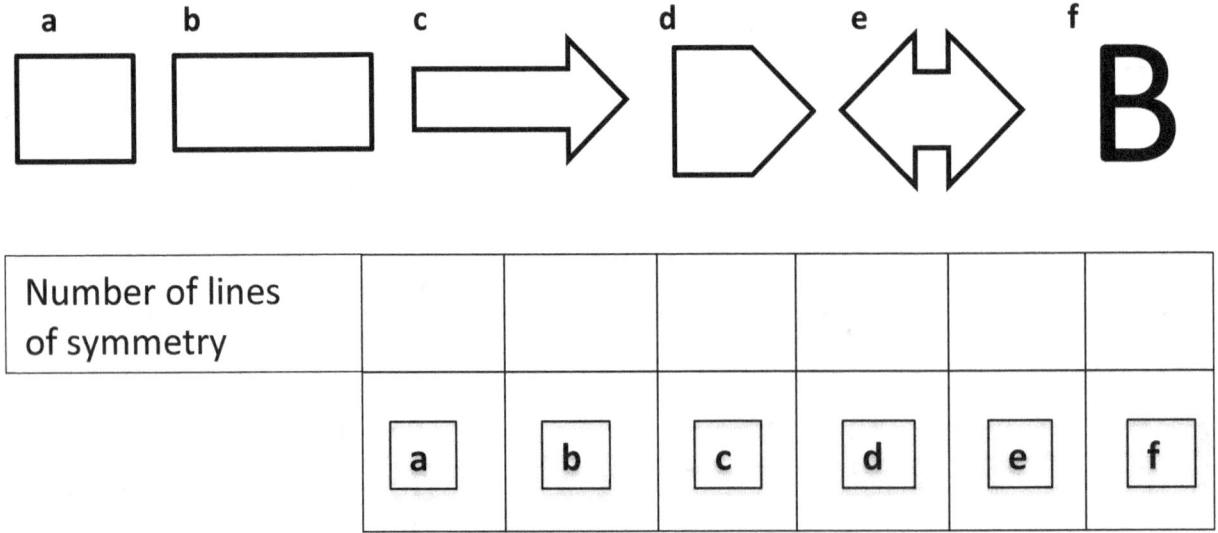

Number of lines of symmetry						
	a	b	c	d	e	f

REFLECTIONS

Some reflections occur in real life, from butterfly to natural occurrences

Can you draw a line of symmetry on each picture above?

1 Reflect (draw) each shape on the other side of the mirror line.

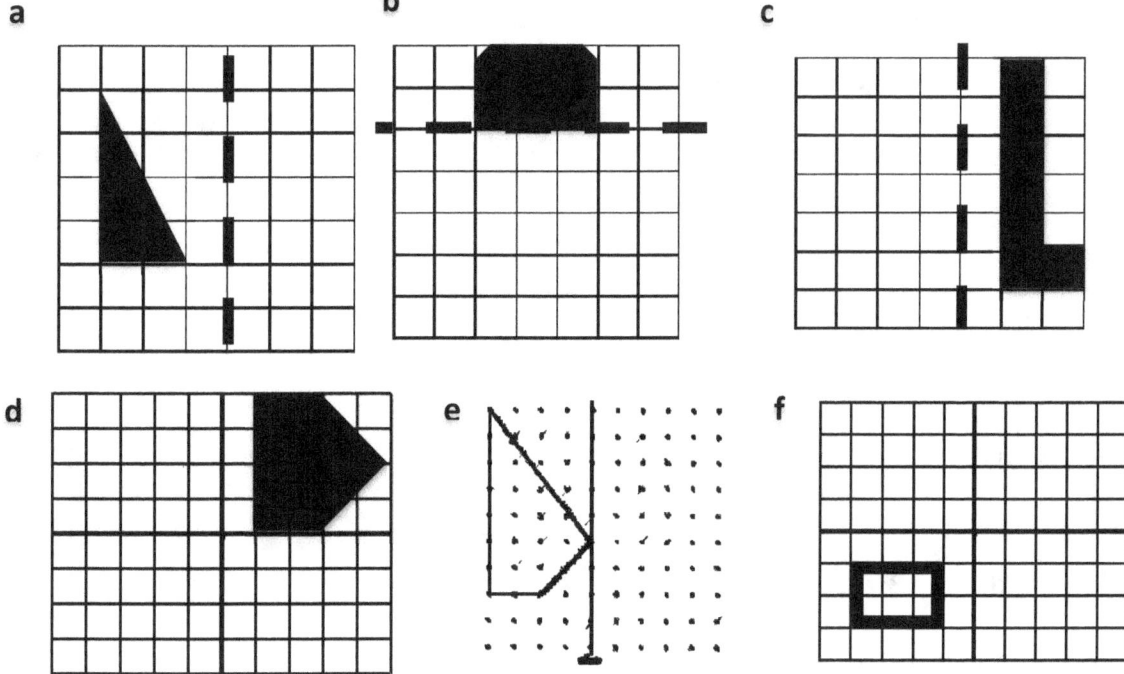

DIVIDING WHOLE NUMBERS

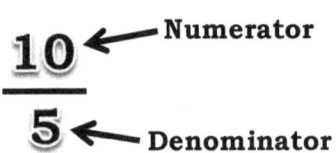

$\dfrac{10}{5}$ ← Numerator / Denominator

$3 \times 5 = 15$
$15 \div 3 = 5$
$15 \div 5 = 3$

Dividing simply means how many of the denominator can you get from the numerator.

$\dfrac{8}{4} = 2$ ✓   because 4 goes into 8 two times or 4 x 2 = 8

1 Write the missing numbers in the boxes provided.

a 8 ÷ 4 = ☐   d 27 ÷ 3 = ☐   g 9 ÷ 3 = ☐

b 10 ÷ 2 = ☐   e 18 ÷ 9 = ☐   h 25 ÷ 5 = ☐

c 18 ÷ 6 = ☐   f 4 ÷ 4 = ☐   i 20 ÷ 2 = ☐

14   18   24   20

2 a) **Pineapple** divided by 2 will give ☐

b) **Strawberry** divided by 9 will give ☐

c) **Pumpkin** divided by 8 will ☐

d) **Apple** divided by 5 will give ☐

COLOURING ACTIVITY 3

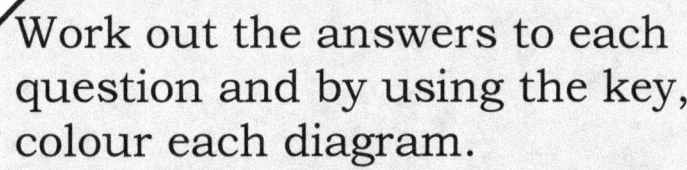

Work out the answers to each question and by using the key, colour each diagram.

Key: 20 = Yellow 10 = Blue
5 = Green 7 = Red

1) 3 + 2 = 5

2) 14 − 4 =

3) 8 + 12 =

4) 10 ÷ 2 =

5) 4 × 5 =

6) 17 − 10 =

7) 2 × 5 =

8) 25 − 5 =

9) 10 × 2 =

10) 19 − 14 =

LENGTHS AND ESTIMATES

1) Estimate the length of each line and measure the exact length in cm.

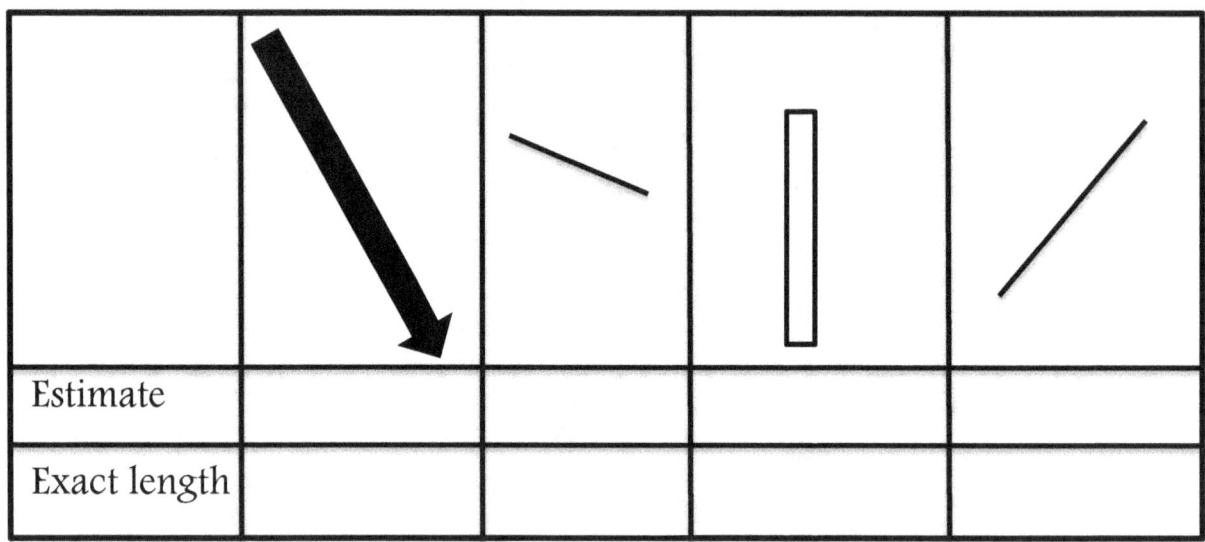

Estimate				
Exact length				

2) Measure and write down the length of each item in centimetre.

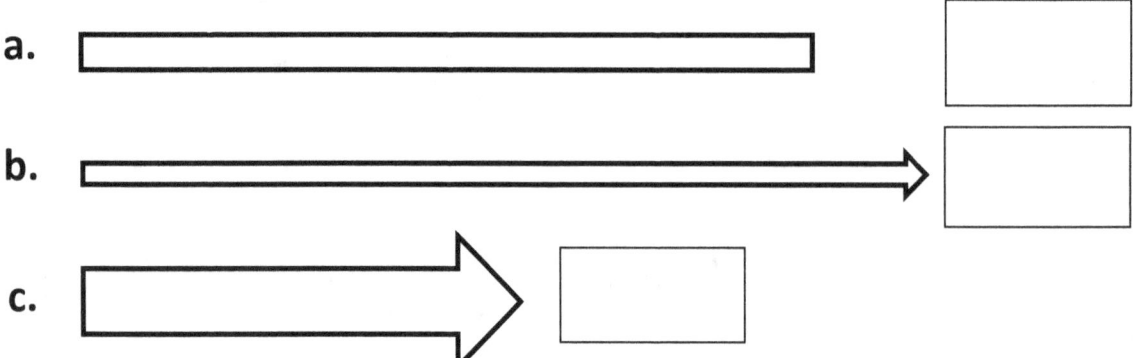

3) Write down the measurements from question 2 above in millimetres.

READING SCALES

1) What is the reading shown on the scale?

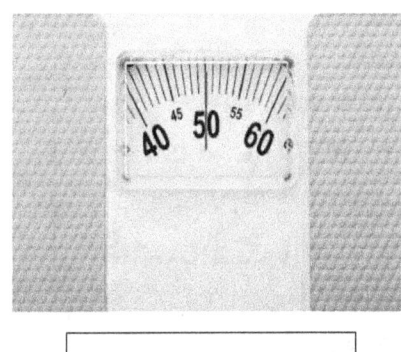

2) What is the reading on the scale?

3) Write down the readings on the scales below.

a
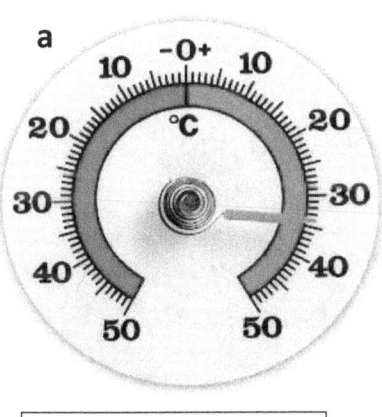

_____ °C

b

_____ kg

c

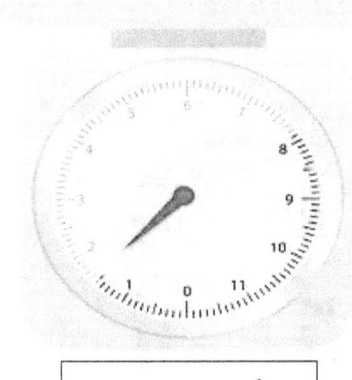
_____ kg

d
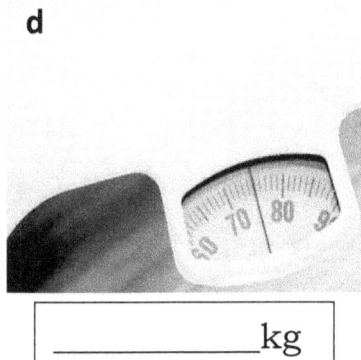
_____ kg

e

_____ °C

f
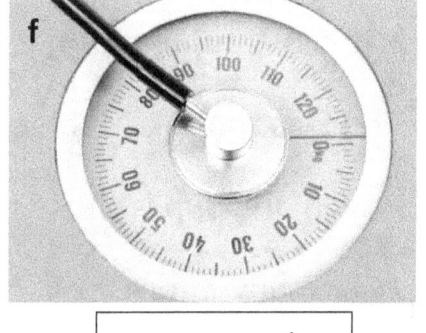
_____ kg

TALLY /BAR CHART

Frequency table		
Colour of pencil	Tally	Frequency
Black		5
Grey	|||| ||	
Blue	|||	

1) Some pupils were surveyed about the colour of pencils used in their classroom every morning. The result is shown on the frequency table.
a Complete the table.
b The favourite colour was ____
c How many pupils were surveyed? [ ]

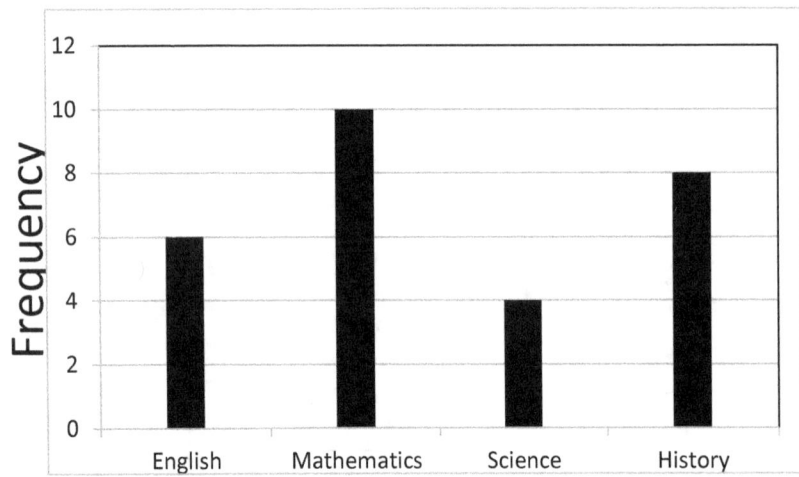

2) A survey was carried out and pupils were asked to choose their favourite subject. The results are shown in the bar chart.

a The most popular subject is ----------------------------------

b The least popular subject is ----------------------------------

c How many pupils chose English? ----------------------------

d How many pupils took part in the survey? ---------------

e Isabel says "half of the pupils surveyed chose English and mathematics." Is she correct? ------------------------------

USING CODES

The numbers below are linked to different shapes.

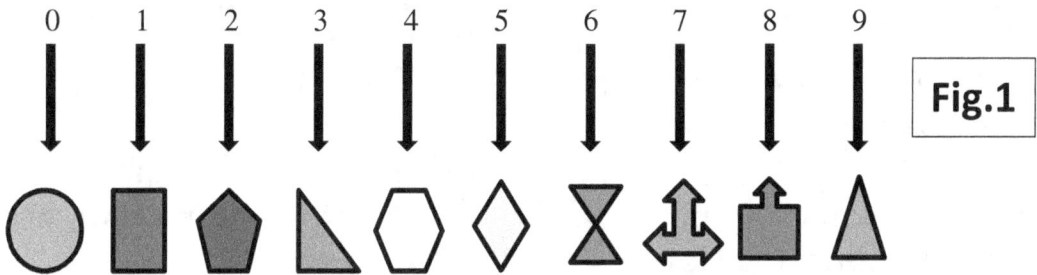

Example: Write down the value of the underlined symbol.

Solution: The symbols represent 4 <u>7</u> 1 0. Therefore, the value of 7 is **700**.
Using the codes and the symbols in fig.1, write down the value of each underlined symbol.

1)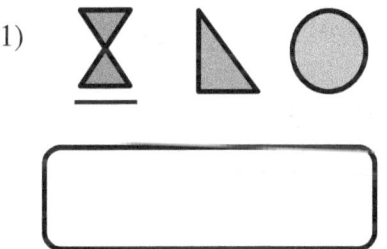

4)

2)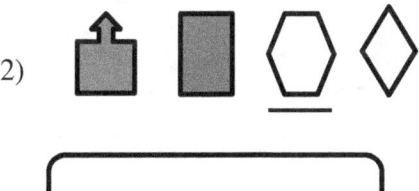

5)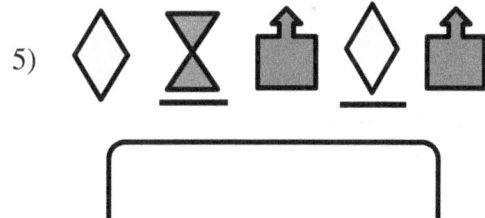

3)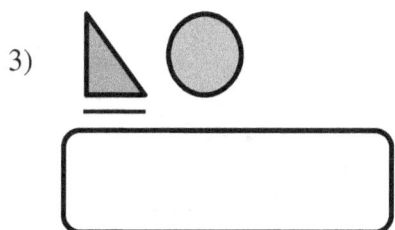

6)

TEST PAPER

* Calculator **not** allowed
* 50 minutes allowed
* 30 marks and above (Excellent)
* 20 – 29 marks (Good)
* 0 – 19 Improvement needed

GOOD LUCK!!!

1 Put a tick on the smaller number from the pair of numbers below.

a) 32   23

b) 2.7   7.2

2 marks

2 Circle **all** the even numbers from the list below.

7   8   20   33   56

2 marks

3 Fill in the missing numbers.

a) $5 + \square = 9$

b) $1 \times 3 = 2 + \square$

c) $8 \div 2 = 2 \times \square$

d) $12 - \square = 4 + 6$

4 marks

4 Write the missing numbers for the patterns below.

a) 2, 4, ___, 8, 10, ___

b) 14, ___, 8, ___, 2

2 marks

5

Happie   Saddie   Bully

a) Happie is **3kg** heavier than Saddie. How much does Happie weigh?

1 mark

b) Saddie, Happie and Bully the dog weighs 30kg altogether. What is the weight of Bully the dog?

2 marks

6 Some cylinders have **odd** and **even** answers. Colour the cylinders using the **key**:

Odd - Blue
Even - White

3 + 5   8 + 3   2 + 12

7 + 0   2 × 7   3 × 3

5 + 5   3 + 12   12 − 6

9 marks

7 $14 + 19 = \square$

2 marks

42

8 From the number lines below, write the numbers marked with letters on the arrow.

a)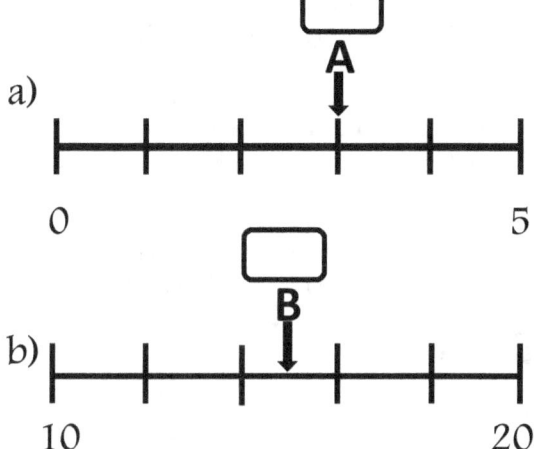

b)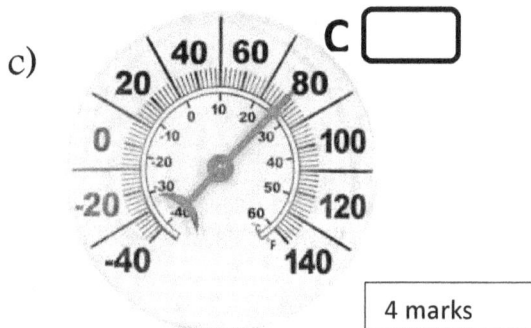

c)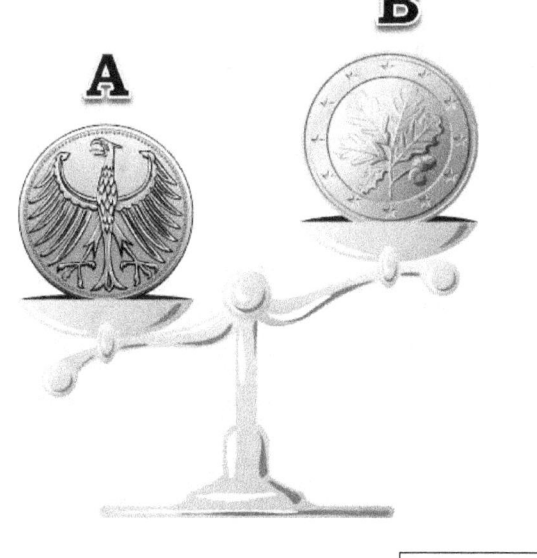

4 marks

9 Which coin is heavier, A or B? Put a tick.

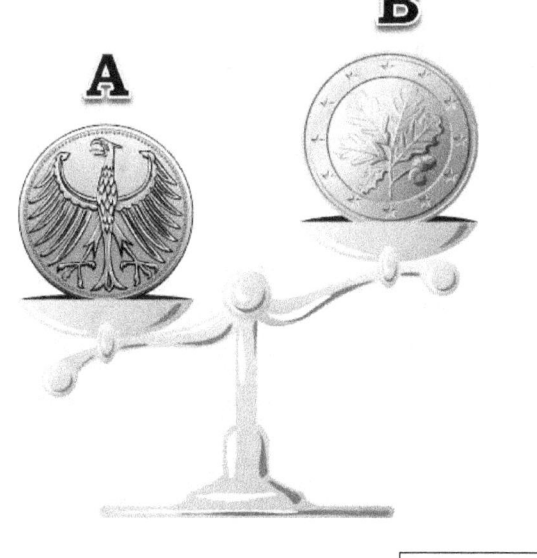

1 mark

10 Starting with the biggest, put these numbers in order of size

| 7 | 6 | 9 | 8 | 4 |

2 marks

11 Look at the shapes below.

Choose from **triangle, rectangle, parallelogram, octagon, circle and hexagon** to complete the following sentences.

a) Shape Q is a _____
b) Shape S is a _____
c) Shape P is a _____
d) Shape R is a _____

4 marks

12 Write the times for P, Q, R and S

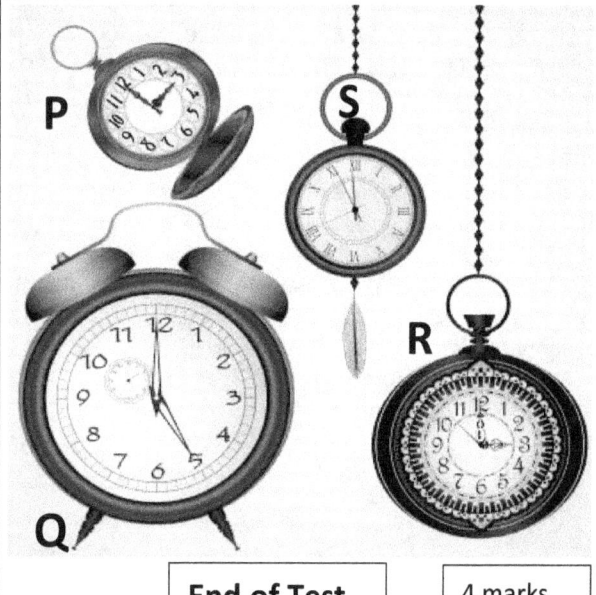

End of Test

4 marks

Answers

Numeracy
1) 6 2) 9 3) 6 4) 18 5) 10 6) 3 7) 6 8) 8
9) 2 10) 1 11) a) 4, 8, 20 b) 1, 5, 11, 17 c) 11, 17, 20
d) 1, 4, 5, 8 e) 1 f) 20 g) 21 h) 9 i) 16 j) 34 k) 32
12) 7 13) 11 14) 9 15) 12 16) 5 17) 14 18) 17 19) 27
20) 15
21)

+	3	5	8	8
9	12	14	17	17
5	8	10	13	13

22) a) 18   b i) 9 + 18 = 27   ii) 27 − 9 = 18

Colouring Activity (1)
1) 10, red 2) 4, green 3) 20, purple 4) 20, purple
5) 4, green 6) 5, brown 7) 7, blue 8) 5, brown
9) 5, brown 10) 7, blue 11) 10, red 12) 4, green

Multiplication
1) 0 2) 2 3) 3 4) 14 5) 18 6) 0 7) 21 8) 32
9) 36 10) 0 11) 2 12) 24 13) 42 14) 35 15) 16
16) 32 17) 18 18) 45 19) 30 20) 100 21) 2 22) 2
23) 2 24) 5 25) 4 26) 2 27) 1 28) 8

Words and Numbers
1) One 2) five 3) ten 4) seventeen 5) twenty
6) 5 – five, 7 – seven, 10 – ten, 15 – fifteen, 20 - twenty
7) 8 8) 12 9) 20 10) 30 11) 35 12) 53 13) 60
14) 66 15) 70 16) 79 17) 80 18) 87 19) 90 20) 100

Even and Odd Numbers
1) 6,8,20 2) 3,7,9,11 3) 20,22,24,26,28, 30 4a) 36 b) 21
c) 15 5) Dad 6) 84 years 7) True 8) 31 9) False
10) Pupils own colouring

Adding Small Numbers
1a) 20 b) 20 c) 25 d) 12 2a) 23 b) 28 c) 143 d) 272
3a) 8 b) 15 c) 12 d) 6 e) 11 f) 9 g) 30 h) 10 i) 17
j) 23 k) 50 l) 40 4) 20 5a) 60 pence b) 40 pence

Place Value
1) 10 2) 10 3) 4 4) 9 5) 30 6) 7 7) 2 8) 40 9) 50
10) 6 11) 2 12) 60 13) 70 14) 8 15) 7 16) 20 17) 100
18) 200 19) 8 20) 300 21) 50 22) 20 23) 5 24) 500
25) 7 26) 10 27) 2 28) 700 29) 60 30) 8000

Missing Numbers
1a) 5,6,7 2) 15,30 3) 13, 17 4) 14, 17 5) 0, 12,15
6) 20, 25, 30 7) 20, 30, 60 8) 8, 20 9) 7, 13
10) 34, 36, 38 11) 20, 32 12) 6, 9 13) 1, 6, 26 14) 16, 20
15) 9, 21 16) 8, 12, 16 17) 13, 31 25, 52 69, 96 123, 321 238, 832 304, 430 234, 432 349, 943 257, 752 123, 321 347, 743 348, 843 115, 511 178, 871

Find My Number
1) 7 11 12 34 13 2) Any combinations of A and B that will equal 40. 3) Any combinations of C and D that will equal 57.

Colouring Activity 2 – Pupils correct colouring using the key.

Small and Big Numbers
1) 12 2) 1 3) 14 4) 35 5) 11 6) 40 7) 50 8) 3.4 9) 8.0
10) 1.1 11) 12.5 12) 4.4 13) 34.4

Number Patterns (Sequences)
1a) 5, 6 b) 10, 12 c) 25, 30 d) 13, 15 e) 12, 10 2a) 7, 9
2b) 20, 50, 2c) 13, 22 2d) 13, 21 e) 80, 82 3) 18,22,26
4) Even 5a) ADD 3 b) ADD 2 c) Take away 3 d) ADD 5
6a) Draws 7 circles   b) 2 c) add 2 each time
7a) 18,28, b) ADD 5 each time 8) take away 5 9a) 3, 5, 7
9b) 7,10,13 9c) 8,6,4 9d) 14,11,8

2D Shapes
1a) ○ b) △ c) ⊕   2a) rectangle b) circle
c) square d) triangle e) pentagon f) hexagon
3) Hexagon ⬡ Rectangle ▭ Triangle △ Pentagon ⬠
Circle ○ Octagon ⬡

3D Shapes
1a) Cylinder b) Cuboid c) Cone d) Pyramid
2a) ▭ b) ○   3) No. Circle is 2D and sphere is 3D.

Symbols and Basic Equations
1a) 6 b) 10 c) 13 d) 13 e) 9 f) 25
2a) 7 b) 0 c) 15 d) 20 e) 20 f) 20 g) 50

Thinking Skills
1a) 50 pence b) £1 c) £4 2a) 30 pence
b) £1.40 c) Yes, 4 rulers will cost £2.80. 3a) 8 b) 4
4a) 56 pence b) i) 1p, ii) 20p, 1p iii) 5p, 1p c) 4 pence

Addition Skills
1a) 47, b) 55 c) 76 d) 98 e) 97 f) 88
2a) 20 b) 80 c) 800 3) 48 4) 95 5) 90 6) 220 7) 86 8) 78 9) 44 10) 94 11) 99

Subtraction Skills
1) 61 2) 22 3) 52 4) 34 5) 82 6) 41 7) 10 8) 10 9) 54
10) 20 11) 7 12) 8

Ordering Numbers
1) 3,4,6,8,10 2) 3,8,9,10,19 3) 4,6,10,16,20
4) 14,24,40,41,56 5) 20,50,56,76,79 6) < 7) < 8) < 9) >
10) > 11) < 12) < 13) >

Time and Clock 1
1) 12 2) 12:30 3) 1
4) 1:30 5) 2 6) 2:30 7) 3 8) 3:30 9) 4 10) 4:30 11) 5
12) 5:30 13) 6 14) 6:30 15) 7 16) 7:30 17) 8 18) 8:30
19) 9 20) 9:30 21) 10 22) 10:30 23) 11 24) 11:30

Time and Clock 2

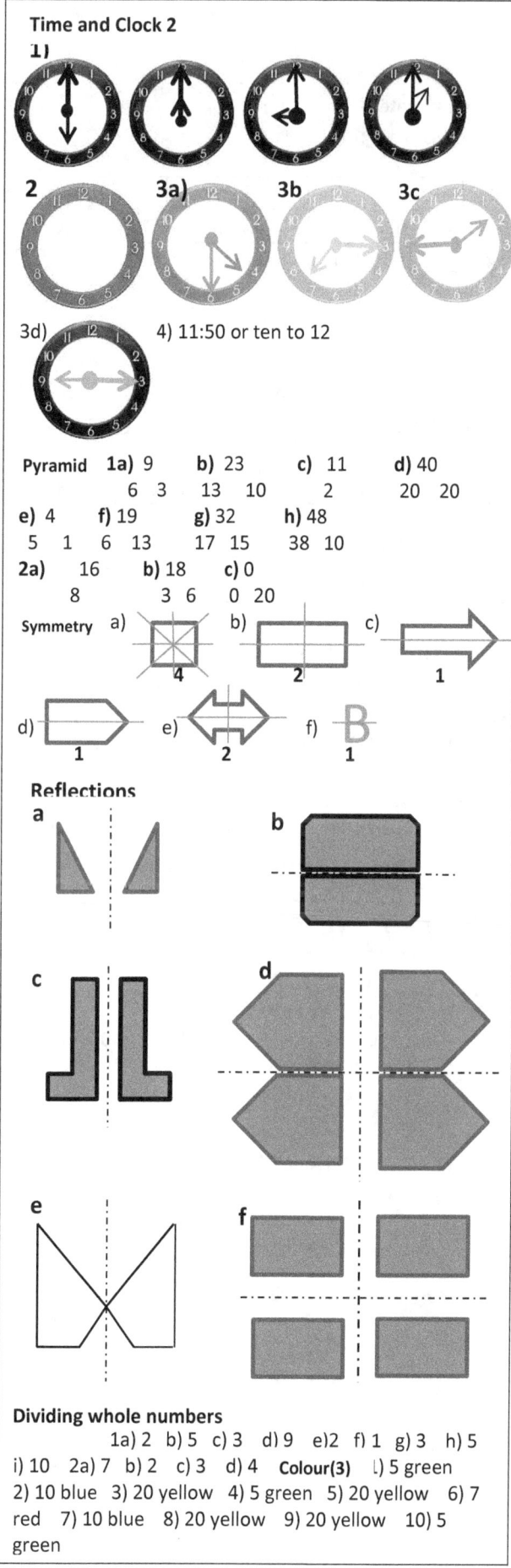

1) (clocks shown)

2 (blank clock) **3a)** **3b** **3c**

3d) **4)** 11:50 or ten to 12

Pyramid **1a)** 9 **b)** 23 **c)** 11 **d)** 40
6 3   13 10   2   20 20
e) 4 **f)** 19 **g)** 32 **h)** 48
5 1   6 13   17 15   38 10
2a) 16 **b)** 18 **c)** 0
8   3 6   0 20

Symmetry a) 4   b) 2   c) 1
d) 1   e) 2   f) 1

Reflections
a, b, c, d, e, f

Dividing whole numbers
1a) 2 b) 5 c) 3 d) 9 e) 2 f) 1 g) 3 h) 5
i) 10 2a) 7 b) 2 c) 3 d) 4 **Colour(3)** 1) 5 green
2) 10 blue 3) 20 yellow 4) 5 green 5) 20 yellow 6) 7 red 7) 10 blue 8) 20 yellow 9) 20 yellow 10) 5 green

Lengths and Estimates
1) All answers ±2mm

Estimate	5cm	1.8cm	3.2cm	2.8cm
Exact L	4.8cm	2cm	3cm	3cm

2) a 10cm **b** 11.5cm **c** 6cm

3) a 100mm **b** 115mm **c** 60mm

Reading Scales
1) 50   2) 44 or 45   3a) 34°C   b) 50 kg   c) 1.5 kg   d) 75 kg   e) 40°C   f) 129 kg

Tally and Bar Charts
1a

Colour	Tally	Frequency
Black	∦	5
Grey	∦ //	7
Blue	///	3

 b Grey   **c** 15 pupils
2 a Mathematics   **b** Science   **c** 6
 d 28   **e** Isabel is wrong. 6 + 10 = 16 and not 14

USING CODES
1) 600   2) 40   3) 30   4) 7000 and 2   5) 6000 and 50   6) 1

ASSESSMENT (TEST)
1a) 23   b) 2.7   **2)** 8,20,56   **3a)** 4   b) 1   c) 2   d) 2   **4a)** 6, 12   b) 11, 5   **5a)** 10 kg   b) 13 kg

6
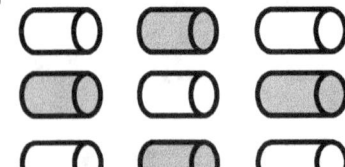

7) 33   **8a)** A = 3   b) B = 15   c) 80   **9)** A   **10)** 9,8,7,6,4   **11a)** Rectangle   b) Circle   c) Hexagon
d) Triangle   **12)** P : 3 o'clock   Q : 5 o'clock   R : 3 o'clock   S) 11:55 or 5 to 12